DISCOVER
Your Career

Careers If You Like Travel

Stuart A. Kallen

ReferencePoint Press

San Diego, CA

ReferencePoint Press®

Printed in the United States

For more information, contact:
ReferencePoint Press, Inc.
PO Box 27779
San Diego, CA 92198
www.ReferencePointPress.com

LIBRARY OF CONGRESS CATALOGING-IN-PUBLICATION DATA

Names: Kallen, Stuart A., 1955- author
Title: Careers if you like travel / by Stuart A. Kallen.
Description: San Diego, CA : ReferencePoint Press, Inc, 2026. | Series: Part of the Discover Your Career series | Includes bibliographical references and index. |
Identifiers: LCCN 2025025004 (print) | LCCN 2025025005 (ebook) | ISBN 9781678212223 lib. bdg. | ISBN 9781678212230 ebook
Subjects: LCSH: Tourism--Vocational guidance--Juvenile literature | Travel--Vocational guidance--Juvenile literature | Hospitality industry--Vocational guidance--Juvenile literature
Classification: LCC G155.5 .K35 2026 (print) | LCC G155.5 (ebook) | DDC 910.23--dc23/eng/20250818
LC record available at https://lccn.loc.gov/2025025004
LC ebook record available at https://lccn.loc.gov/2025025005

Contents

Introduction: Earning on the Go

Many dream of earning a living while traveling the world. They envision working jobs that allow them to visit historic sites and museums, lounge on beaches, and experience diverse cultures. And some travel-friendly careers do provide perks that can satisfy a worker's wanderlust. Some travel bloggers, photographers, fashion models, and musicians get to join the jet set and see the world. But these fields are overflowing with freelancers who struggle to make a living. As travel blogger Caroline Rowe explains, "If you're considering trying to become a travel blogger, bear in mind that there are apparently over 600 million blogs on the internet and less than 5% of them make money."[1]

Rowe points out that creating a successful travel blog takes a great deal of time, effort, and skill. But there are more stable careers that include both travel opportunities and paychecks. Cruise ship lines consistently hire restaurant staff, housekeepers, casino workers, deckhands, fitness instructors, security officers, entertainers, and others. These workers earn money, travel the world, and receive free room and board. Those who are employed as tour guides by international travel companies obtain similar benefits.

Beyond the tourism industry, there are numerous other careers that offer workers a chance to travel. The business travel management company Expensify reports that US business travelers take over 405 million long-haul business trips in an average year. More than 60 percent add one to four extra vacation days to these trips. Those who are employed in sales, construction, engineering, nursing, and teaching can take advantage of this combination of business and leisure, what Expensify blogger Shaelyn Combs calls "bleisure."[2] While bleisure is not for everyone, some people—like world-traveling computer consultant Alex Walling—do their best work while they are on the go. "I love working from coffee shops, flights, Uber rides, hotels, anywhere that has me moving around in new and interesting environments," Walling writes. "I also think that being really busy forces me to prioritize what I'm working on and adds some excitement to the job."[3]

Not a Vacation

Walling points out that business travel is not the same as taking a vacation. And like many others who travel for a living, he admits that life on the road, in the air, and on the sea can be difficult. Those who travel for business are usually on tight schedules that require them to attend numerous meetings, client dinners, conferences, and other work-related events in a short time. They need to be mentally sharp and focused on the tasks at hand. And business travelers often contend with severe weather, jet lag, broken rental cars, lost luggage, and bad food. According to statistics compiled by Expensify, more than half of all those who traveled for work in 2024 experienced delayed or canceled flights, and over a third expressed travel-related anxiety about their safety and security.

Those who work in the travel industry have a different set of problems. Airline and cruise ship employees are often required to put in twelve-hour days. When they have time off, they are often too tired to go sightseeing or club-hopping. And many who work for typical travel industry wages do not have extra funds to go out to eat or party in expensive tourist destinations. Traveling for work can also be lonely. Wellness blogger Sejal remarks, "If you're away for long periods at a time, there could be homesickness to contend with. And if you've got a family, you may have to get used to spending less time with them."[4]

Whatever the drawbacks, many are eager to take jobs that allow them to travel. And opportunities abound. The World Travel &

Tourism Council says the global travel and tourism industry supported 348 million jobs worldwide in 2024. Not all these jobs provide travel prospects to employees. But airlines, hotels, and other industry players offer free or discounted travel vouchers to employees. And Expensify says nearly six in ten corporations have policies that allow workers aged twenty to thirty to combine business trips with vacation time. Such policies are aimed at a new generation of workers who do not wish to spend their days in office cubicles. They built their careers to blend travel and work and are setting their sights on perks that provide them with plenty of bleisure.

What Does an English Teacher Abroad Do?

After graduating college with a degree in speech communications, Erika Copeland had one objective: "My goal was to travel the world and live abroad for as long as possible,"[5] she says. Copeland landed a job teaching English to fifth graders in Incheon, South Korea. She did not speak Korean, but those who work as English teachers abroad do not need to speak the language of their host country. While some English teachers abroad are bilingual or multilingual, it is not usually a job requirement.

At a Glance

Number of Jobs
250,000

Pay
$57,683 in 2024*

Educational Requirements
Bachelor's degree

Personal Qualities
Fluent English speaker, good communicator, adaptable, organized, creative

Working Conditions
Forty hours a week, indoors in classrooms

Future Job Outlook
Growth of 3 percent through 2029

* For all English teachers

Copeland's one-year assignment in South Korea ended in 2010, but she was not ready to return to the United States. She took an online course in Teaching English as a Foreign Language (TEFL). This course enabled Copeland to pursue job opportunities across Europe. During the following decade, she taught English to students in Germany, France, and the Republic of Georgia. Copeland says she deeply values the connections she made with others while teaching abroad: "You never feel alone in your profession. . . . I have friends from all over the world now that are teachers or former teachers. We have a very deep connection because of our experiences working in foreign countries with foreign students in foreign classrooms. These experiences really bring people together."[6]

In 2024 Copeland was working with students in the Republic of Georgia while running an international online business that delivered English lessons to adults. She was able to build her career as an English teacher abroad because there is a great demand for those who work in this profession. The statistics website Statista says around 1.5 billion people in the world speak English. This makes English the most popular language on earth. But only around one-quarter of those speak English as their first language. The rest speak English as a second language, referred to as ESL. Those who teach ESL devise lesson plans, hand out assignments, give tests, and document student progress. The teachers do not focus on formal grammar and word usage but teach what is called conversational English. This is basic, casual speech, including slang words, that people use in everyday communications.

In addition to teaching skills, those who work in foreign countries must understand the complex steps that are necessary to obtain visas and work permits. English teacher Mark Williams, who has taught in Thailand, China, and South Korea, says this can be frustrating and time-consuming: "If there's one thing that has constantly tested my patience, my resilience, and my sanity, it's visas and work permits. . . . I didn't picture myself standing in long immigration lines, filling out endless forms, or having minor clerical errors put my entire career on hold."[7]

A Typical Workday

Those who teach English abroad often work in classrooms that are little different from those found in the United States. Because it is easier for younger children to learn a second language, English teachers abroad are most often assigned to elementary schools, where they teach students in grades one through five. Teachers

Be Eager and Flexible

"Most schools will ask for a [detailed résumé], a cover letter, a recent photo of you and two references. Once you have everything prepared, you can send out blind applications. . . . If you're still home, also add a projected date for when you'll be in [the country]. Don't expect an email back if you're not selected for an interview. If you haven't heard from a school, politely follow up to request an interview. Talk up any experience you have; if you haven't taught before, express your eagerness to learn and your flexibility—these are two attributes I look for in new teachers."

—Cat Gaa, English language teacher in Spain

Quoted in Nomadic Matt, "How to Teach English and Live in Spain," August 27, 2024. www.nomadicmatt.com.

might have up to six different classes a day, each lasting from forty minutes to one hour. Lessons are often based on themes that resonate with students, such as TV shows, music, food, family, or kid-friendly topics like dinosaurs and space travel.

Many teachers use games and other activities to make learning fun. One of the most popular games is called Board Race. Students are divided into teams. A word like *animal* is written on a whiteboard, and students take turns writing a type of animal underneath. Each correct English word is worth one point; misspelled or wrong words are not counted. The team with the most points wins. Other games concentrate on speaking, listening, word order, spelling, comprehension, and vocabulary. As English language teacher Emma Lander writes, "Whether you're teaching adults or children, games will liven up your lesson and ensure that your students will leave the classroom wanting more. Games to learn English can be used to warm up the class before you begin, during the lesson to give students a break when you're tackling a tough subject, or at the end of class when you have a few minutes left."[8]

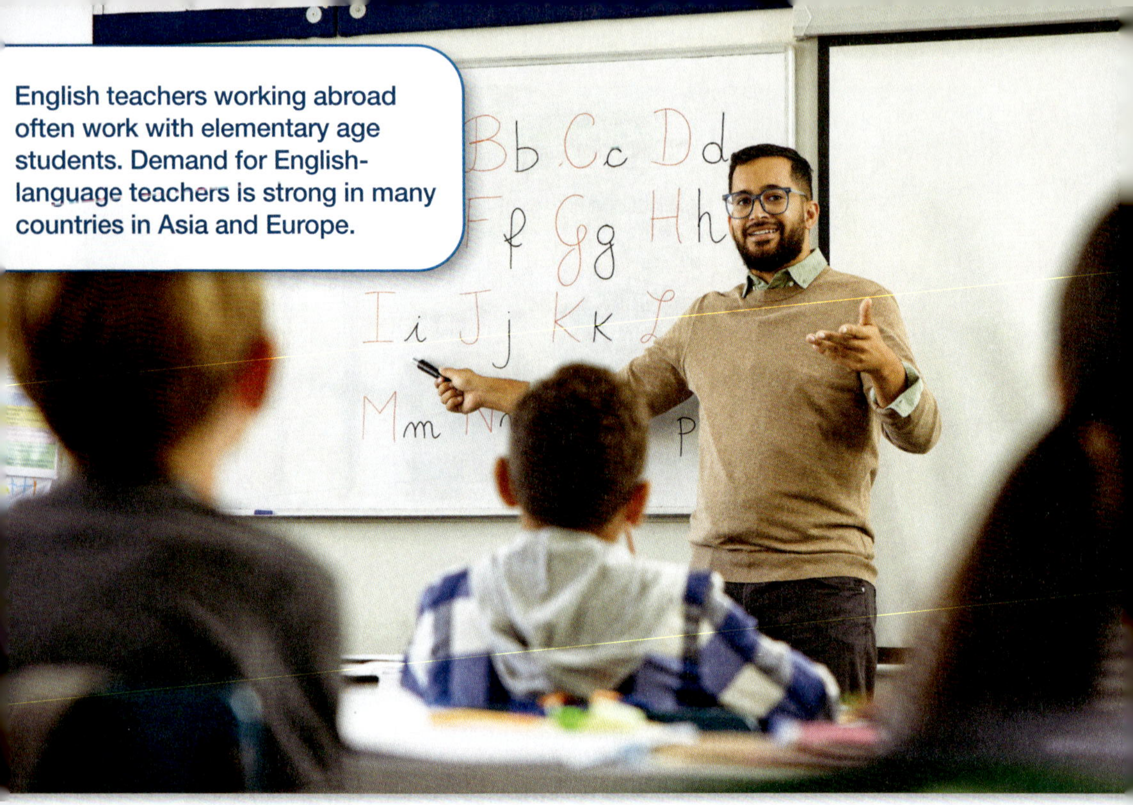

English teachers working abroad often work with elementary age students. Demand for English-language teachers is strong in many countries in Asia and Europe.

When they are not working directly with students, teachers have other responsibilities. They advocate for student needs, meet with parents and officials, and work with other teachers to coordinate lesson plans. Teachers are also in charge of planning English-only events for students, including plays, concerts, and film screenings that mix learning and entertainment.

Education and Training

The requirements for teaching English abroad vary from country to country, but most nations expect teachers to hold at least a bachelor's degree in any subject. Those who hope to make a career as an English teacher abroad often take college courses in English, linguistics, and English as a second language. Students might also want to learn Spanish, Mandarin, or any other language that is spoken where they hope to teach someday. Courses that highlight adult education help prospective teachers learn about teaching people from various cultures and backgrounds.

Certification is required in most countries, including China, Japan, South Korea, and Spain. English language teachers must have a certificate in either TEFL, TESOL (Teaching English to Speakers of Other Languages), or CELTA (Certificate in Teaching English to Speakers of Other Languages). While these acronyms can be confusing, those who want to teach English abroad should understand what they mean. The certifications can be obtained online or in person. Depending on the type of certification, course length can range from 40 to more than 150 hours and generally takes up to three months to complete. Subjects include language learning, linguistics, classroom management, lesson planning, student evaluation, and cultural sensitivity in teaching. Students need to pass a series of quizzes and tests to obtain certification.

There are formal documents required for teaching abroad. A teacher must possess a valid passport and then obtain a work visa issued by the nation where the teacher will be employed. Most countries also conduct criminal background checks on applicants.

Skills and Personality

Those who successfully teach English abroad rely on the same skills that any teacher would use in a classroom. They are patient, have a passion for teaching, and are good communicators who clearly speak fluent English. Being multilingual is a bonus, especially when it comes to understanding the native language of the host nation. Not only does this make it easier to communicate with students, it also helps when shopping, going out to eat, traveling, and making new friends.

Creativity is a bonus for teachers, who need to develop innovative lesson plans, games, and events that engage students. Teachers are also highly organized; they need to keep order in the classroom and provide lessons logically and efficiently.

Those who teach English abroad do not see themselves as tourists and do not view their work as an extended vacation.

Teachers have many responsibilities that require them to be hard workers. Accommodations are rarely luxurious, and home comforts like air-conditioning and good Wi-Fi might be lacking. This means teachers need to be flexible when adapting to new cultures.

Workplace Settings

Teachers abroad generally work twenty to thirty hours a week in a classroom, with extra time spent grading papers and creating lesson plans. But conditions vary from country to country. In Japan teachers sometimes work from 9:00 a.m. to 10:00 p.m., with students coming in after the regular school day. Teachers are also expected to work some weekends at school events.

Employers and Pay

English language teachers work in public schools, colleges, international schools, training centers, language academies, and online. Some work as tutors for one or more students.

The employment website Indeed says English teachers in the United States earn an average annual salary of $57,683. But salary for English language teachers abroad varies from nation to nation. Those who work in France can expect to earn an average of $24,000 annually, with few extra benefits. English language teachers in South Korea earn an average annual salary of $25,800. But they also receive valuable benefits, including free housing, paid vacations, health insurance, and flight reimbursement. English language teacher William Davies says China is one of the best places to work: "The demand for English teachers is high in China, with a wide variety of TEFL jobs scattered across the country. . . . Teachers in China can benefit from a high salary and benefits, an excellent quality of life, and valuable work experience."[9] Davies says English teachers could expect to earn up to

Listen to Your Students

"If you're willing to listen and observe, your students will give you lesson ideas. . . . [By] paying attention to what they write about in their assignments, you can incorporate their interests into lesson plans. Youth culture in Japan is unique and a perfect opportunity to pique your students' interest in talking about these trends in English. This can save you time or even get the creative juices flowing when you're stuck on ideas for creating lesson plans that will engage students, helping them learn and participate in a dynamic atmosphere. You may think you're there to inspire the students, but they will inspire you, too!"

—Kimi Sugiyama, English teacher in Japan

Kimi Sugiyama, "The Pros and Cons of Teaching English in Japan," Go Overseas. July 15, 2024. www.gooverseas.com.

$3,000 a month at public schools in China in 2025, while some international schools pay up to $5,000 monthly. As in South Korea, teachers in China receive housing, airfare, health insurance, and other perks.

What Is the Future Outlook for English Teachers Abroad?

The job market for all English teachers is expected to increase by around 3 percent through 2029 However, the International TEFL Academy estimates that there are around 250,000 people teaching English abroad. They work in around forty thousand schools and language institutes throughout the world. The academy says that around half of these teachers leave their positions and return to their home country every year. There is also a growing demand for online instructors. This means that those who can teach conversational English to people in other nations can expect to find rewarding work in nations around the globe.

Find Out More

Bridge

https://bridge.edu

Bridge is a language education organization that places certified English language teachers in over 140 different countries. Bridge offers training, professional certification, and job opportunities.

Council on International Educational Exchange (CIEE)

www.ciee.org

CIEE is a nonprofit organization that promotes international education and exchange. The Accrediting Council for Continuing Education & Training certification provided by CIEE is respected all over the world and never expires. The website offers job search assistance and informational blogs.

TEFL Org

www.tefl.org

This organization that focuses on teaching English as a foreign language provides TEFL certification, employment resources, online teaching opportunities, and blogs about teaching in other cultures.

What Does a Flight Attendant Do?

In February 2025 a commercial jetliner crash-landed at Pearson International Airport in Toronto. By the time the jet came to rest on the snowy runway, it had flipped upside down and burst into flames. Seventy-six passengers were left hanging by their seat belts from the inverted floor of the aircraft. Despite the disorienting shock of the impact, the two flight attendants on board immediately jumped into action, while ground crews rushed to extinguish the flames. The attendants shouted out evacuation instructions loudly and clearly as they freed people from their seats and hurried them toward the exit doors. Videos posted on social media show the flight attendants acting in a calm, firm manner, ordering passengers to put their phones down and leave their luggage behind. Flight attendants are trained to evacuate jetliners within ninety seconds, and despite the craft being upside down, the crew in Toronto managed to do just that. While several passengers were critically injured, no one died.

At a Glance

Number of Jobs
130,300

Pay
$68,370 in 2024

Educational Requirements
High school diploma

Personal Qualities
Customer service skills, good communicator, physical strength, attention to detail

Working Conditions
World travel, standing and walking for long periods, lifting luggage, possible life-threatening emergencies

Future Job Outlook
Growth of 10 percent through 2033

Due to privacy concerns, the names of the flight attendants were never publicly released. But according to aviation crash investigator Graham Braithwaite, they were heroes. "These people put their lives on the line—they're the last people off the airplane, and I think sometimes we forget that," Braithwaite said. "They serve us drink and food, and that's wonderful, but their real function is to keep you safe."[10]

Flying is one of the safest modes of travel, and most flight attendants never have to deal with an emergency like the one in Toronto. But the flying public often forgets Braithwaite's main point. Travelers see flight attendants in their crisp uniforms greeting passengers, making sure carry-on bags are safely stowed, demonstrating safety procedures before takeoff, and serving passengers during flights. Unknown to many, airlines are required by law to train flight attendants to evacuate airplanes and administer and coordinate emergency medical care. Flight attendants learn to deal with fires, crashes, terrorist attacks, and even drunk and unruly passengers.

Anyone who works as a flight attendant will explain that the job has distinct positive and negative qualities. On the positive side, flight attendants are world travelers. They spend layovers (off-duty hours between flights) in world-class locales. International flight attendants regularly visit cities in the United States, Europe, Asia, South and Central America, and Australia. Even when they are not on duty, flight attendants—and their families—can fly for free or at deeply discounted rates. Another perk concerns flexible schedules. Flight attendants with several years of seniority can pick and choose the days and nights they wish to fly. They also get to make friends all over the world and sometimes meet interesting people, including business titans, politicians, and celebrities.

The negative aspects of working as a flight attendant vary depending on how long the person has been working for the

Finding Fascinating Friends

"One of my favorite things about flying is the people I get to work with. I find them interesting and diverse. Some have no college degree, while others have PhD's. Some grew up in Europe, Thailand, or South America while some have never really moved away from their hometown city in the United States. The people you meet will inspire and fascinate you. . . . I'm lucky to have friends in so many cities across the globe because I've been a flight attendant and I've chosen to travel. Having friends to meet up with, wherever you may be, is a big pro to cabin crew life."

—Kara Watts, flight attendant

Kara Watts, "Pros and Cons of Being a Commercial Flight Attendant," Flight Attendant Life, 2025. https://flightattendantlife.com.

airline. Those who have worked for more than a few years have greater choices when it comes to picking destinations and schedules. Newer hires have little say about when and where they fly, and their schedules can be unpredictable from one week to the next.

Jet lag affects almost every air traveler, and flight attendants are no different. Jet lag occurs when a traveler crosses multiple time zones and their natural sleep–wake cycles are disturbed. When it is midnight in Los Angeles, it is 8:00 a.m. in London, England, and it can take more than a week for the body of a California flyer to adjust to the difference. Most flight attendants are not in one place that long, and jet lag can cause sleep disruption, brain fog, anxiety, irritability, and headaches. As a flight attendant named Kara Watts writes, "Time zones and jet lag will play a mean number on you. This is unavoidable and one of the biggest downsides to working as a flight attendant. I find jetlag to be hell. I went through periods during my commercial aviation career where I was so fatigued that I could not sleep for more than two hours at a time."[11]

A Typical Workday

On a typical day, flight attendants are at work long before passengers board a plane. Ninety minutes before takeoff on an international flight, they are in a briefing room with pilots going over flight plans, weather forecasts, and possible security issues. After the briefing, flight attendants inspect emergency equipment to ensure it is working.

Flight attendants work in pressurized airplane cabins, serving passengers drinks and food. When meals are served, they heat and prepare dishes and ensure passenger dietary restrictions are followed. Flight attendants make safety announcements on the aircraft's PA system, including advising passengers about turbulence, seat belt usage, and other incidents. When the flight is over, flight attendants write reports about safety, medical, or security issues that may need attention.

Education and Training

Flight attendants must be high school graduates. While a college degree is not necessary, competition for flight attendant jobs is fierce. Applicants with a degree in hospitality, tourism, public relations, or communications have a better chance of qualifying for an interview with an airline. Language proficiency is a plus for those who hope to work on international flights.

The Federal Aviation Administration (FAA) requires flight attendants to train for up to eight weeks in flight regulations, first aid, and emergency procedures like airplane evacuation. Upon completion of training, the FAA awards flight attendants a Flight Attendant Certificate of Demonstrated Proficiency. Flight attendants are required to update their certificate periodically throughout their careers.

Those who work as flight attendants need a valid passport and must pass drug tests and criminal background checks. They are

The job of a flight attendant includes greeting passengers, making sure carry-on bags are safely stowed, demonstrating safety procedures before takeoff, and serving passengers during flights. They are also trained to respond to emergencies.

required to have vision that is correctable to at least 20/40 and often need to conform to height requirements set by the airline. Flight attendants may also have to regularly pass a medical evaluation.

Skills and Personality

Flight attendants need to be physically fit because they spend most of their workdays standing and walking around an aircraft. They often stoop to help children and bend and lift to retrieve meal trays. Attendants stretch and reach while handling baggage that can weigh up to 45 pounds (20.4 kg). Full beverage carts can weigh up to 250 pounds (113.4 kg), and flight attendants are required to push and pull them up and down the slanted aisles of

Many Benefits

"You have to like dealing with the public. You have to be able to handle situations that arise with calm and professionalism. People come in all sorts and some will be super nice, others will be total pains. . . . [But if] you love to explore new places and people, that helps a lot as your time off can be very enjoyable and educational depending on what airline you work for. Flying domestic isn't as thrilling as flying international, but you still get benefits and that may help you to explore the world."

—Debbi Fuller, flight attendant

Debbi Fuller, "What's It like Being a Flight Attendant?," Quora, 2024. www.quora.com.

the aircraft. During emergencies flight attendants might need to lift injured passengers. As former flight attendant Jennifer Pierce notes, "Flight attendants are workers who perform an abundance of manual labor, and without experiencing the job duties firsthand, it's hard to measure how much physical exertion it requires."[12]

Flight attendants need excellent customer service skills to interact with hundreds of people every day in very close quarters. Travelers can be irritable, tired, and rude, and flight attendants need to address their needs with patience, poise, and tact. Good communication skills are a must for flight attendants, who clearly provide directions and safety instructions in person and through the plane's PA system. Flight attendants also need to be decisive and alert, aware of security and safety risks during a flight while possessing the ability to remain calm and make good decisions in emergencies.

Workplace Settings

Airlines operate around the clock, and flight attendants are often required to work nights, weekends, and holidays. FAA safety regulations mandate that flight attendants maintain a schedule where they work an average of twelve to fourteen days a month. They can expect to spend sixty-five to eighty-five hours a month in the air, not including overtime. Those who are assigned to international flights might work up to eighteen hours in a single day. Attendants typically spend another fifty hours a month on the ground preparing flights, writing reports, and waiting for planes to take off. Schedules vary, but two-day trips are the most common for flight attendants; they complete one flight, have a layover of around ten hours, then return to their home base or fly to another city. Airlines pay for food and accommodations for those who stay overnight.

Flight attendants experience a higher-than-average number of work-related injuries caused by lifting luggage, experiencing turbulence, and being exposed to passengers' contagious diseases.

Employers and Pay

Flight attendants work for major airline companies like Delta Air Lines, Southwest Airlines, and American Airlines. Some work for smaller, regional airlines while others are employed by charter companies, which are private aviation firms that cater to business executives and other wealthy individuals.

The Bureau of Labor Statistics (BLS) says the median annual wage for flight attendants was $68,370 in 2024. The highest-paid 10 percent of earners made more than $104,100.

What Is the Future Outlook for Flight Attendants?

The demand for air travel continues to increase every year. Due to this trend, the BLS predicts employment for flight attendants will increase by 10 percent through 2033. As the travel industry booms, flight attendants will be needed to ensure the safety and comfort of airline passengers.

Find Out More

Association of Flight Attendants-CWA (AFA-CWA)

www.afacwa.org

The AFA-CWA is a flight attendant labor union that negotiates for better pay, benefits, and working conditions for its fifty thousand members who work at twenty airlines. Students visiting its website can learn about scholarships and find information about many issues flight attendants face.

Association of Professional Flight Attendants (APFA)

www.apfa.org

The APFA is a labor union for flight attendants that publishes the latest industry news, media campaigns, and political actions taken by its members. Its website provides a window into the world of flight attendants.

Federal Aviation Administration (FAA)

www.faa.gov

The FAA regulates all airlines in the United States. Its website describes the requirements to obtain a Flight Attendant Certificate of Demonstrated Proficiency, as well as pertinent information such as the FAA medical and security requirements and preemployment tests.

What Does a Construction Manager Do?

The University of Texas at San Antonio (UTSA) offers degrees in construction science and management (CSM). At graduation ceremonies, those who complete the program wear construction hard hats instead of the traditional mortarboards. And every year the number of bright orange hard hats crossing the stage at UTSA is growing. In 2024 there were more than three hundred students enrolled in CSM, which was once seen as a minor degree program. This trend is visible throughout the country. According to the National Student Clearinghouse Research Center, the number of student enrollments in construction programs has been growing by more than 19 percent annually since 2023, faster than science and technology or business management. The associate professor who leads the UTSA program, Ibukun Awolusi, explains why growing numbers of students are majoring in CSM every year: "A lot of our students have a full-time job opportunity before they even finish their degrees. We have 100% placement. You're not going to get that in a lot of [other] programs."[13]

At a Glance

Number of Jobs
520,900

Pay
$104,900 in 2024

Educational Requirements
Bachelor's degree

Personal Qualities
Good communicator, knowledge of construction methods, business skills, leadership qualities

Working Conditions
Full time with overtime, travel required

Future Job Outlook
Growth of 9 percent through 2033

Not only are job opportunities for CSM graduates numerous, they also pay well compared to other jobs in

Bringing Designs to Life

"Let's think about a typical building. The architects will design the building . . . the civil engineers will do the structural design. . . . But who does the actual construction? Construction scientists and managers. We take all of those designs, and we estimate the cost of all the resources, including materials needed to bring the building to life. We have to make sure those materials are properly sourced, which can be a challenge when there are supply chain issues. We schedule when items will be purchased and avoid cost overrun. We schedule the labor and supervise them to make sure they perform each task safely."

—Ibukun Awolusi, associate professor of construction science and management

Quoted in Shae Conner, "Q&A: Professor Explains Why Students Flock to UTSA for Construction Management Careers," UTSA Today, October 16, 2024. www.utsa.edu.

the building trades. Most companies promise large starting salaries for construction managers, who can begin their careers earning six-figure incomes.

Construction management attracts entrepreneurial, hardworking individuals who possess the skills to oversee large construction projects. Sometimes referred to as general contractors or project managers, construction managers plan, coordinate, and supervise construction projects large and small. They act as the bridge between clients who are paying construction costs and the architects, engineers, contractors, and workers who complete the project. Construction managers prepare cost estimates for an undertaking, create working budgets, and put together schedules for various phases of a project. They have a strong grasp of construction techniques, the costs of a wide range of building materials, and complex building codes. Another major priority for construction managers is preventing accidents and ensuring that crews have a safe work environment. They are required to analyze the safety of jobsites, enforce safety regulations, and conduct safety meetings. Construction managers also institute safety protocols when necessary.

A Typical Workday

The work of a construction manager continually evolves over the course of a project, which means there is no typical day. When at their office, construction managers spend a great deal of time talking on the phone and exchanging texts and emails with dozens of others who are working on a project. As building trades journalist Celeste Gable writes:

> One of the most critical roles of a construction manager is coordination. They act as the central hub of communication between different teams, including architects, engineers, contractors, and subcontractors. They manage the flow of information, resolve conflicts, and adapt to unforeseen challenges. Problem-solving is a daily exercise, as construction rarely goes entirely as planned.[14]

The job requires managers to produce hundreds of pages of documents with software like Procore or Autodesk Construction Cloud. The programs are used to create bidding documents for upcoming projects, manage budgets and schedules, assign resources to tasks, schedule deliveries, analyze safety and building regulations, and track job progress.

Construction managers need to be on-site to coordinate building processes with carpenters, masons, ironworkers, and other trade workers. Construction managers also provide supervision over specialized processes, such as excavation, paving, or landscaping. They study blueprints and plans with workers to ensure everyone on a jobsite knows what to do and when to do it. Construction

Construction managers plan, coordinate, and supervise construction projects. They interact with clients, architects, engineers, contractors, and workers.

managers who oversee multiple projects in a large geographic region might travel extensively to building sites in different cities or states. Those managing big projects like airports or hospitals might be required to temporarily move to a new location. Because of the travel involved, a construction manager's "jobsite office" is often the front seat of a pickup truck, where laptops, tablets, and cell phones are used to run cost estimates, deal with emergencies, and execute hundreds of other objectives that the fast-paced job requires.

On large projects, a construction manager might work with a team of junior managers who handle various aspects of a project. For example, one junior manager might oversee the pouring of a building's concrete foundation, while another might coordinate electrical work or plumbing.

Education and Training

Some construction managers work their way up to the job. They are promoted to oversee projects after many years working as carpenters or electricians. However, construction projects have become more complex over the years, with new building techniques, automated equipment, and stricter environmental and safety requirements. This makes it increasingly important for prospective construction managers to obtain a bachelor's degree in construction science, construction management, architecture, or construction engineering.

Dozens of colleges and universities offer CSM degrees. Courses include project control and management, cost estimation, building codes, worker safety regulations, construction materials and design, and business administration. Students study mathematics, statistics, and other relevant subjects.

Most who major in construction management programs get hands-on experience working as interns. Because of the strong demand for construction managers, internships are often paid, unlike those in many other fields. The employment website ZipRecruiter says construction management internships in California paid $93,922 annually in 2025. Those who work as interns solicit bids from subcontractors, schedule workers, learn new technologies and concepts, and interact with clients, regulators, and financial personnel. Oftentimes, interns are hired as full-time employees after they graduate.

Certification is not mandatory to become a construction manager. However, those who seek accreditation demonstrate to prospective employers that they are willing to invest time and effort in expanding their knowledge. Organizations such as the Construction Management Association of America and the American Institute of Constructors offer certifications that can be obtained by completing self-study courses and passing exams.

Skills and Personality

Construction management is a demanding job that requires commitment, competence, and dedication to excellence. Construction managers are self-starters; they get up early and get things done. They make quick decisions on the fly to keep projects on schedule and have the communication skills required to give orders, smooth out difficulties with clients, and explain complicated tasks to workers. Construction manager Adam McNair writes, "[Construction managers need] the ability to communicate clearly and concisely, applying appropriate grammar, listening attentively, presenting ideas in a logical, compelling manner in both small and large group settings. . . . Communicating technical information to a non-technical audience [is also important]."[15] Construction managers also need a good background in construction techniques, technical skills to interpret blueprints, and business acumen to manage the money side of the business.

Construction managers rely on their analytical skills to solve the inevitable problems that arise on a regular basis over the course of any building project. They need to patiently deal with delayed deliveries, bad weather, erratic employees, and demands from building inspectors.

Those who manage building projects are team leaders, and supervising large crews can be difficult, according to construction manager John Hansen: "The hardest part of the job is managing people not managing construction. You will need to become a business manager, diplomat, nursemaid, camp counselor, and referee."[16]

Managing people, heavy workloads, tight deadlines, big budgets, safety issues, and building regulations can be extremely stressful. Construction managers must balance demands from harried clients and government inspectors who can shut down a project if it does not meet strict codes and safety requirements. Construction managers are also ultimately responsible for accidents, missed deadlines, and cost overruns. They have to explain what went wrong and how they will fix it. Successful construction managers are experts at putting aside personal feelings and acting in a calm and rational manner no matter the situation.

Having Fun While Learning

"Every day I check emails to see if we've been awarded a contract to do a job we bid on. If we get the contract, we get the plans and start the estimating process. We go to the job site and get measurements. We assess our team to do the job and put on more people if needed. I'm able to read blueprints, inquire about materials, and order them. . . . [I decide] what we will need for a project and which team members will be needed first. I track our progress . . . and I'm learning a lot. . . . It's just really fun."

—Yesenia Murillo, construction manager

Quoted in San Joaquin Valley College, "Grad Q&A with Construction Management Graduate Yesenia Murillo," 2025. www.sjvc.edu.

Workplace Settings

Construction managers work full time with plenty of overtime in offices and at building sites. As with most other workers in the building trades, construction managers are early risers. They are often on jobsites at dawn, and they take care of problems before heading into the office for meetings with clients, government regulators, and others. And there is little downtime, as construction managers often work on several projects at the same time.

Employers and Pay

Construction managers work for real estate companies, property developers, construction contractors, civil engineering firms, government agencies, and builders of all sorts of structures. According to the Bureau of Labor Statistics (BLS), construction managers earned an average annual salary of $104,900 in 2024. Those who manage commercial construction projects such as shopping centers, office buildings, and other nonresidential projects earned around $6,000 more. Construction managers who worked for residential home builders earned about $92,980.

What Is the Future Outlook for Construction Managers?

The BLS says employment of construction managers is expected to grow by 9 percent through 2033. Construction managers will be needed to oversee the construction of hospitals, schools, office buildings, retail outlets, apartments, homes, and other structures. Those who have construction experience combined with degrees in construction science, construction management, and civil engineering are expected to have the best job prospects.

Find Out More

American Council for Construction Education (ACCE)

www.acce-hq.org

The ACCE promotes and accredits quality construction education programs. Visitors to its website can find lists of ACCE-approved colleges and universities and find resources for certification, training, careers support, and scholarships.

American Institute of Constructors (AIC)

https://aic-builds.org

The AIC is a society of construction professionals that provides certification for construction managers and others. The institute provides online educational courses and webinars, and the AIC Career Centers offers job postings and professional development advice.

Construction Management Association of America (CMAA)

www.cmaanet.org

The CMAA offers in-person training, online learning courses, mentor programs, and certification for construction managers. Visitors to its website can contact advisors, access articles and white papers, and reference the association's library.

What Does a Travel Nurse Do?

Jennifer Traub was twenty-eight years old when she became a registered nurse (RN) in 2017. Traub had never journeyed far beyond her home in Florida before she took a job as a travel nurse in Southern California. Travel nurses are registered nurses who work temporary short-term assignments at various locations. These assignments range from two to twenty-six weeks, but thirteen weeks is most common. They fill vacancies in hospitals, nursing homes, children's hospitals, community clinics, and special care facilities. Traub was very nervous about moving across the country to unfamiliar surroundings but did not let her fears stop her. In 2025 she wrote that signing on as a travel nurse was the best thing she ever did. "I have worked in almost 30 different hospitals all over the states of California and Florida. I have [worked in] 33 countries on 4 different continents," she said. "As much as I have seen, I know there is still so much more I have not. I have made a valiant attempt, however, to satiate my burning desire to learn as much as I possibly can."[17] Most travel nurses do not journey as extensively as

At a Glance

Number of Jobs
1,733,502

Pay
$101,132 in 2024

Educational Requirements
Bachelor's degree

Personal Qualities
Hardworking, dependable, good communication skills, empathetic, emotional strength

Working Conditions
Twelve-hour shifts, three days a week in hospital settings

Future Job Outlook
Growth of 6 percent through 2033*

* For all registered nurses

Traub. But those in her profession have the flexibility to travel as much or as little as they desire while earning a good paycheck wherever they land.

Travel nurses perform the same tasks as all registered nurses. They examine patients to assess their problems, make records of patient symptoms and medical histories, administer shots, provide medicine, and change dressings on wounds. Nurses assist doctors with procedures, operate medical equipment, perform diagnostic tests, and help patients manage their illnesses and injuries.

Many travel nurses are specialists. Cardiovascular nurses specialize in caring for patients who have heart disease. They aid doctors in operating rooms during heart surgery. Critical care nurses work with seriously ill or injured patients. Emergency room nurses treat those who are suffering from life-threatening conditions in hospital emergency departments. Neonatal nurses administer to the needs of newborns and mothers. As RN Debra Wood writes, "Travel nursing covers a variety of specialties, including medical–surgical and specialty units, such as the emergency department, operating room, labor and delivery, or intensive care unit."[18]

A Typical Workday

Most travel nurses work three twelve-hour shifts each week, which allows time to see sights and explore new settings. When they first arrive in a new workplace, they go through an orientation period that lasts a few days. During this time travel nurses learn the layout of the hospital. They meet staff and learn standard procedures. They might undergo training to operate unfamiliar equipment or learn about a facility's electronic records system.

Travel nurses bring their own stethoscopes and uniforms called scrubs, which need to match the colors of those worn by others in the facility. Most other equipment is provided. Like other nurses, travel nurses spend their shifts caring for patients, assisting doctors, providing patient education, and documenting the

An Adaptable Attitude Is Vital

"The secret to succeeding as a travel nurse is adaptability. By adopting flexibility, you can avoid burnout and approach every task with positivity and an open mind, which makes it simpler to overcome obstacles and take pleasure in novel experiences. . . . Having a positive outlook can change your experience by helping you see change as an exciting opportunity rather than a stressful situation. . . . You can handle the highs and lows of travel nursing with poise and confidence if you practice gratitude and resilience, which ultimately contributes to your sense of stability."

—Emma Joyce, travel nurse

Emma Joyce, "Comfort on the Go: Creative Ways to Feel at Home as a Travel Nurse," Gypsy Nurse, April 14, 2025. www.thegypsynurse.com.

care that is provided. As Wood writes, the fun begins when the shift ends: "Travel nurses can use their time off to explore their new city, try local cuisine, visit museums, attend the theater, or go to the beach or on a hike in a local park. They might connect with other travel nurses or local friends."[19]

Education and Training

Travel nurses have the same educational background as registered nurses. They either have a two-year associate's degree in nursing or a four-year bachelor of science in nursing. Most staffing agencies that place travel nurses prefer those with a bachelor's degree because the higher level of education is useful to those who often work in diverse and ever-changing environments.

High school students who hope to become travel nurses can prepare for their future careers by taking courses in biology, chemistry, statistics, and if possible, anatomy and physiology. English and writing courses are also a plus since colleges prefer students with good communication skills.

Travel nurses work temporary, short-term assignments at locations nationwide. They fill vacancies in hospitals, nursing homes, community clinics, and specialty care facilities.

High school students can also obtain certification in first aid and basic life support by taking classes offered by the Red Cross and the American Heart Association. Health care facilities are always seeking volunteers and often have programs for high school students. Nursing schools often have more applicants than they have available spots. Volunteer nursing hours are always a plus for students hoping stand out on college applications.

Those who graduate from college are required to obtain an RN license in the state where they intend to practice. This requires them to pass the National Council Licensure Examination for Registered Nurses. The in-person exam is said to be challenging, but in 2024 over 93 percent passed on their first try. Those who fail can take the exam again after a waiting period.

Skills and Personality

Travel nurses are adaptable to change, a skill that is necessary when working with new people in new settings. Exceptional communication skills are vital, because mistakes can be dangerous or even deadly in a hospital setting. Travel nurses need to speak and write clearly to ensure that instructions they give to others are properly carried out. Travel nurse recruiter Scott Barry writes:

> As your shift comes to an end, your focus will adjust to wrapping up and setting the next shift up for success. This may include leaving detailed handoff notes if another traveler or temp is coming in or touching base with the charge nurse to ensure continuity of care. Closing out the day as a travel nurse means being extra diligent in communication to bridge the gap between temporary and permanent staff.[20]

Listening skills are also important when attending to patients and following directions from doctors and other staff. Travel nurses need to be dependable, patient, and flexible and require good leadership qualities to solve problems and resolve conflicts.

Like all nurses, travel nurses are hard workers. They see many patients every day, fill out numerous forms, and are on their feet for hours at a time. Travel nurses need to be physically fit and should understand how to practice self-care, including eating properly and getting enough sleep. Travel nurse Kristina B. explains:

> Adequate rest is essential for physical and mental health. Travel nurses may have irregular schedules, but they should aim for a consistent sleep routine when possible. Use blackout curtains, earplugs and white noise machines to create a conducive sleep environment. Maintain a balanced diet even while on the move. Plan meals in advance, choose healthy options, and avoid excessive caffeine and sugary snacks.[21]

Travel nurses need to be emotionally strong to deal with the reality of the job. Registered nurse Kelita Raphael says, "One of the toughest parts of this profession is realizing that, though we truly try our best to provide life-saving care, not everyone can be

Expect the Unexpected

"As a travel nurse, my day varies. Usually I arrive at work, take reports, and round on my patients to introduce myself, ensure they're okay, and answer any questions they may have that came up during change of shift. I verify that any fluids or drips are correct, monitors are connected and functioning, etc. . . . I maintain constant communication with the provider in regard to any issues or recommendations, and I monitor for changes in my patients that require further intervention. Although this may be a basic day, each day is different, and you honestly never know what to expect when working as a nurse."

—Kelita Raphael, registered nurse

Kelita Raphael, "A Day in the Life of a Travel Nurse," ATI, 2025. https://atinursingblog.com.

saved. Our patients suffer debilities and even lose their lives. For me, that's one of the most heartbreaking parts of being a nurse."[22]

Workplace Settings

Most travel nurse jobs are found in what are called acute health care facilities that include emergency rooms, intensive care units, and specialized care units. Acute facilities deal with accidents, heart attacks, at-risk newborns, and other urgent medical conditions. Like other nurses, travel nurses work a variety of shifts, including days, evenings, nights, and overnight. They may work weekends and holidays. Shifts tend to last twelve hours, but some might be shorter or longer.

Travel nurses are generally provided with perks that include housing stipends or free company-arranged housing, reimbursements for travel, and bonuses. Travel nurses can choose to work in big cities, suburbs, or rural settings, and many select places with warm weather or natural beauty.

Employers and Pay

Most travel nurses sign contracts with staffing agencies that recruit nurses and place them in clinics, hospitals, and other health care facilities throughout the world. Staffing agencies provide a list of employment opportunities that match the nurse's qualifications and preferences.

Regarding pay, travel nurse Katie Dixon writes, "A travel nurse's salary is considered a standout benefit of the job, as it's usually higher than that of a staff nurse. Hospitals and health-care facilities leverage high pay to attract travel nurses to facilities struggling with staffing shortages."[23] While compensation varies from place to place, the job pays an average of 25 percent more than the median wage for RNs overall. According to the employment website ZipRecruiter, the average annual pay for a travel nurse in 2024 was $101,132. Top earners in the field were bringing in over $132,500.

What Is the Future Outlook for Travel Nurses?

The Bureau of Labor Statistics (BLS) says job growth for all registered nurses will increase by 6 percent through 2033. The BLS does not have specific statistics for travel nurses. However, the demand for travel nurses grew by 430 percent from 2018 to 2024, according to travel nurse agency AB Staffing, which expects the trend to continue as the United States experiences an increasingly aging population.

Find Out More

American Society of Registered Nurses (ASRN)

www.asrn.org

The ASRN represents the interests of registered nurses in the United States. The group provides information about scholarships and student grants and provides access to the *Journal of Nursing* and other publications.

National Council Licensure Examination (NCLEX)

www.nclex.com

NCLEX is the premier licensing organization for registered nurses. Its website contains information about certification, including qualification, preparations, exams, scoring, and results.

Travel Nurse Across America (TNAA)

https://tnaa.com

The website of TNAA provides comprehensive information about travel, benefits, and pay for travel nurses. Students can find out how to become an RN and read information about certification and job placement.

Sales Engineer

What Does a Sales Engineer Do?

Millions of people work in sales, while countless others are employed as engineers. Some, however, utilize aspects of both jobs and work as sales engineers. These professionals use their understanding of technical, scientific, and engineering concepts to sell complex products and solutions to business customers and the government.

At a Glance

Number of Jobs
59,700

Pay
$121,530 in 2024

Educational Requirements
Bachelor's degree

Personal Qualities
Math skills, highly organized, interpersonal skills, analytical

Working Conditions
Full time with overtime in offices and on factory floors, extensive travel

Future Job Outlook
Growth of 6 percent through 2033

Sales engineers are first and foremost engineers, professionals who are trained to design, test, and create components of manufactured items. Most have a background in electrical, computer, mechanical, or industrial engineering. Depending on their experience, sales engineers might specialize in software products, robotic machinery, medical devices, manufacturing equipment, or other complicated systems that require specialized knowledge.

Sales engineers might be thought of as human versions of owner's manuals that come with automobiles and other consumer products. Sales engineers have a comprehensive knowledge of the products they are selling and can convey their function to customers who might not be experts. Pete Robertson, who moved from working as a data center engineer to sales engineering, explains the job:

> There's an educational side of selling, helping people understand technology, understand how it works, what it's actually achieving. And so I wanted to do that for my customers. I wanted customers to understand what technology could do for their business. And ultimately, customers are buying technology not for technology's sake, but because it's enabling them to do something. And I liked bridging that gap of explaining how technology was addressing a non-technical need.[24]

Salespeople who work with the public might try to convince a customer that their product will make the customer look good or feel better. Sales engineers sell to chief technical officers, engineers, and other educated buyers who cannot be swayed by generalities. Sales engineers must know all about industry standards and regulations related to their products. They will have to present statistics and hard data to clients to convince them to spend large sums of money to purchase new equipment.

Because of the costs and complexities involved in purchasing highly technical systems, companies rarely make quick purchases. Sales engineers have to draw up detailed proposals that might be presented to a company's board members, who vote to decide whether to make the purchase. In many cases a sales engineer can spend up to a year working with a client before a sale is approved. After the sale is closed, sales engineers oversee installation and upkeep of the machinery and resolve problems that might arise.

Long-distance business travel is often a major part of the job for sales engineers. Tech journalist Andrey Melnichuk states, "En-

Preparing to Demonstrate a Product for a Client

"Before facing the client, thorough preparation [for a demonstration] is essential. This starts with a thorough understanding of the client's needs. Thoroughly research their industry, company size, and specific challenges; this will allow you to tailor the demo to their requirements. It is advisable to have preliminary conversations with the client to identify their goals and expectations. . . . Preparing the demo environment is another crucial aspect. . . . Test all technical aspects to avoid mishaps during the demo. Gather supporting materials, such as slides, videos, success stories, and testimonials, that will enrich the experience and provide tangible evidence of the value you offer."

—Luis Armando Vasquez Ruiz, sales engineer

Luis Armando Vasquez Ruiz, "Master the Art of the Demo," North American Association of Sales Engineers, November 19, 2024. https://sales-engineering.org.

gineers may spend a significant amount of time traveling to meet customers at their manufacturing locations. . . . If you do not like traveling, this job is not for you. Sales engineers may visit clients once a week or even more frequently. And if the geographic spread of your customers is extensive (different countries and even continents), supporting them can be quite difficult."[25]

A Typical Workday

Many of the tasks a sales engineer performs in a day are like those of any salesperson. Mornings are spent scheduling meetings with potential customers, checking on deals in progress, and troubleshooting problems for current clients. Sales engineers are usually team leaders who work daily with sales reps demonstrating new products, setting goals, and strategizing methods for reaching new customers. Meetings might also be held with personnel who focus on marketing, product development, and engineering.

During the afternoon, sales engineers might be visiting clients or conducting virtual meetings to demonstrate their products and explain their services. Business management consultant Yang Lu describes this aspect of the job for software sales engineers: "A sales engineer would typically discover what the problems are that a business faces and via a demonstration show how his/her software can mitigate it and paint the wonderful rosy picture of what the software can do for the prospect."[26]

Sales engineers answer technical questions at meetings or with texts, emails, and phone calls. When meetings are over, they will write detailed technical proposals for prospective customers that include implementation plans and cost breakdowns. As the day ends, sales engineers will check back with customers. They might also spend time conducting research into new products, industry trends, and the offerings of competitors.

Education and Training

Anyone wishing to work as a sales engineer needs at least a bachelor of science degree in mechanical, manufacturing, electrical, or industrial engineering. High school students who wish to pursue a career in engineering should focus on STEM subjects—science, technology, engineering, and math. Computer coding skills are very important since engineers commonly work with C++, Python, Java, and other programming languages. Beyond technical skills, engineers rely on teamwork—communications classes that focus on speech and writing can help students learn to express their ideas more effectively.

College students working toward a bachelor's degree in engineering take courses that focus on pneumatics, hydraulics, logic, microprocessors, and integrated systems. Some engineers work

Keep the Customers Satisfied

"If the customer buys, we've only won half the battle. . . . It is my job to remove any obstacles the customer might have to using our product. If they are using the product, I can use the time to do a mini-discovery. I can take the time to see if they need anything else. If so, work on that opportunity while making sure they are happy with our current solution. If they don't need anything else at that time, at least they are happy using our product."

—Ramzi Marjaba, sales engineer, security software

Ramzi Marjaba, "A Week in a Life of a Sales Engineer," We the Sales Engineers, 2025. https://wethesalesengineers.com.

toward a bachelor's degree in one field, such as mechanical engineering, and a master's degree in another field, like electronic engineering. Those who hope to specialize in sales might take courses in business and marketing.

Students in their sophomore year can apply for cooperative education programs (co-ops) at tech companies or at government research labs. Students in co-ops stop taking classes to work full time. The positions are typically paid and last anywhere from three to twelve months. Students receive academic credit and a letter grade for their paid work experience.

Most sales engineers are accredited as professional engineers (PE). To become certified, an engineer must pass a Fundamentals of Engineering exam and work under a licensed PE for four years before taking a Principles and Practice of Engineering exam.

Skills and Personality

Almost all engineering is based on math, and sales engineers need a strong foundation in algebra, geometry, probability, trigonometry, calculus, and statistics.

Sales engineers often juggle numerous projects that are in different phases of completion. This requires them to be highly organized. Sales engineer Ramzi Marjaba writes:

> Keep in mind that no matter what you do, your life will be some sort of chaos. I, for one, do enjoy ORGANIZED chaos. Therefore I try to organize my week as much as possible. I aim to break down my days into specific repeatable tasks. For example, on this day, at this hour, I visit some customers. It's not always a success, as I may show up and the customer is not available, but at least they know I was there to help, and I know what I'm doing in my week.[27]

Like all salespeople, sales engineers need to be outgoing and amiable, with practiced social skills that allow them to make conversation with strangers. As Pete Robertson says:

> I'm not saying that introverts do not have a place in sales engineering at all, but you do need to put yourself in a situation where you're comfortable presenting in front of an audience, where you're comfortable being uncomfortable. . . . You've got to be comfortable managing a room, figuring out in an audience who's the person that's taking too much of the attention away from where you're trying to lead the conversation.[28]

Analytical skills are a must for sales engineers to be able to study a potential customer's needs and tailor solutions to their problems. Business skills are necessary in order to explain the many ways a business can profit by purchasing their product.

Workplace Settings

Sales engineers work in business-to-business or business-to-government sectors. They work full time in offices or on factory floors, and they travel anywhere their employer's products are in use. When big deals are brewing, sales engineers can put in twelve-hour days. Those who work in manufacturing might be on call all day and need to be available in the middle of the night when equipment malfunctions or breaks down.

Employers and Pay

The Bureau of Labor Statistics (BLS) says around 20 percent of sales engineers are employed in the manufacturing sector. Equal numbers work for merchant wholesalers and companies that design computer systems. Around 6 percent are employed by software publishers. Wherever they work, sales engineers can earn good salaries. The BLS says the average annual pay for sales engineers in 2024 was $121,530. Sales engineers who worked for software publishers earned $137,650. Those who worked for merchandise wholesalers were at the lower end of the pay scale, earning an average of $109,150.

What Is the Future Outlook for Sales Engineers?

The BLS predicts that job growth for sales engineers will grow by 6 percent through 2033, slightly faster than the average for all occupations. As digital and manufacturing products become more technologically sophisticated, there will be a continuing need for sales engineers.

Find Out More

Manufacturers' Representatives Educational Research Foundation (MRERF)

https://mrerf.org

The MRERF provides educational programs, curriculum, resource material, and certification for those who work as manufacturer representatives, including sales engineers. The foundation also hosts conferences and funds research.

National Society of Sales Engineers (NSSE)
www.nationalsse.org
The NSSE provides information for aspiring sales engineers and those who work in the field. The society sponsors university organizations on numerous college campuses across the country and offers mentorship programs and student volunteering opportunities.

North American Association of Sales Engineers (NAASE)
https://sales-engineering.org
This organization is dedicated to advancing the role and visibility of sales engineers while offering coaching services to those in the field. The association also offers the NAASE Certified Sales Engineer credential to signify expertise and contributions to the field.

What Does a Commercial Pilot Do?

When commercial pilots stride through airports in their crisp uniforms, they command respect. That respect is well deserved because commercial pilots are responsible for navigating massive aircraft through the skies at speeds of nearly 600 miles per hour (966 kph) while safely transporting passengers or cargo. Commercial pilots commonly refer to their jobs as "pushing the metal," but most of their work takes place on the ground. A few hours before takeoff, a pilot will download the flight plan and take care of preflight checks. Pilot Rida Rais enumerates some of these routines: "We check details like passenger count, baggage loading, aircraft weight, and fuel required for the journey from our base to the destination. Additionally, we review the weather forecasts for departure, destination, and alternate airports, assess the aircraft's status, and perform calculations for its takeoff and landing performance at the destination."[29] These tasks mean that airline pilots often work twelve or

At a Glance

Number of Jobs
152,800

Pay
$198,100 in 2024

Educational Requirements
Bachelor's degree, one thousand-plus hours of flight training

Personal Qualities
Communication skills, quick reaction time, observational abilities, problem-solving skills

Working Conditions
Full time in airports and airplane cockpits, one hundred hours flight time per month

Future Job Outlook
Growth of 5 percent through 2033

more hours a day even if they spend only a few hours at the controls of an aircraft.

Pilots must also periodically undergo wellness and aptitude tests. For example, they are required to regularly see doctors for physical exams and pass annual reviews to remain in compliance with Federal Aviation Administration (FAA) regulations. The pilots' rigorous routines help make air travel the safest mode of transportation.

Depending on their employers, commercial pilots operate a variety of aircraft, from small propeller airplanes and helicopters to large corporate aircraft, cargo planes, and passenger jets. All conduct preflight checks on aircraft equipment and fill out flight logs. When planning a route, they must understand how their aircraft will perform given its weight, power, and speed. These details help pilots determine fuel use and anticipate how to properly navigate through or avoid potential weather hazards.

A Typical Workday

Most of what airline pilots do takes place behind closed doors, either in the restricted flight operations area of the airport or inside a locked cockpit. They hold short briefings with cabin crews to update emergency procedures, receive reports on malfunctioning equipment, and discuss expected delays and weather issues. In the air, pilots are aided by computerized autopilot systems that keep an aircraft aloft without human direction. This allows pilots to take meals, check charts, or do other tasks while the plane is flying level and straight. But an autopilot cannot replace a human operator in maneuvers like taking off, climbing, descending, and landing while simultaneously dealing with weather conditions and avoiding other aircraft. If there is severe turbulence, a pilot might deviate from the original route and adjust the aircraft's altitude to obtain a smoother flight. Therefore, human experience and foresight are necessary to ensure safe and efficient flights.

Life of a Cargo Pilot

"As a cargo pilot, any type of schedule you can imagine is available—domestic, international, 1-day trips, 14-day trips, all-daytime flying, all-nighttime flying, and every combination of these. It is a myth that we only fly in the middle of the night! We fly a lot of U.S. mail, and therefore, have an extensive day flight schedule in the United States. . . . At my airline, each pilot has a company-issued iPad and can access the Flight Planning tools . . . at any time. As a cargo pilot, you can find every type of flying schedule in a variety of aircraft. There is something for everyone here!"

—Kandy Bernskoetter, cargo pilot

Kandy Bernskoetter, "A Day in the Life of a Cargo Pilot," Cleared to Dream, 2025. https://clearedtodream.org.

Besides autopilot computers, commercial jet pilots are assisted in their duties by human copilots, called first officers, who share flight duties such as communicating with air traffic controllers and monitoring engines and other aircraft systems during flights.

Education and Training

About one-third of the commercial pilots in the United States have military experience, according to *The Washington Post*. Most of these pilots learned to fly in the military. Those who are not military veterans have invested a substantial amount of time and money in flight school to become licensed commercial pilots.

Most airlines only hire pilots with a bachelor's degree in any subject. In addition, those who wish to become commercial pilots must take flight training from certified instructors at flight school. Pilot training is offered by some colleges and universities as part of two- and four-year aviation degree programs. Flight instruction can cost hundreds of dollars an hour, and experts say a student

In addition to flying their planes to destinations near and far, commercial pilots plan routes, do preflight checks on aircraft equipment, and log all flight activity. Ultimate responsibility for the safety of passengers and crew rests with the pilot.

wishing to obtain a commercial pilot rating can spend more than $150,000 on flight courses and in-air training.

Anyone wishing to obtain a commercial pilot's license from the FAA must accumulate 1,000 hours of flight time. Helicopter training requires another 150 hours of flight time. Additionally, commercial pilots need a certification called an instrument rating that allows them to fly through clouds, storms, and other limited-visibility conditions while relying on their instruments. An instrument rating is also required to carry passengers on flights of more than 50 miles (80 km) or at night. Commercial pilots are required to pass the FAA Practical Test, informally called a check ride, each year. During this exam, pilots fly in an aircraft with an authorized examiner to demonstrate their skills and competence. Pilots can lose their jobs if they do not pass the check ride.

Passenger airline pilots are required to have an Airline Transport Pilot certificate. This certificate expands various privileges granted by the commercial pilot's license. Most who work for legacy airlines also receive what is called an aircraft-type rating for each model of jet they fly. For example, an individual might earn an aircraft-type rating for an Airbus A320, a Boeing 737, or a Boeing 777.

Beyond the paperwork and certifications, pilots need to be in good physical condition because people's lives depend on their abilities. Captains need to pass a thorough physical exam every six months; other commercial pilots need to pass an annual exam. Health problems such as heart disease, high blood pressure, or failing eyesight might end a commercial pilot's career.

Skills and Personality

The ability to make decisions under pressure is an important skill for pilots. They need to quickly and calmly respond to weather extremes, turbulence, equipment failure, and emergency situations. And pilots need strong technical proficiency to understand the array of gauges, levers, switches, and other equipment in a cockpit. Flight instructor Alec Myers writes:

> The airline pilot gets to operate some of the biggest, fastest and most expensive civilian machinery that exists. . . . In order to get to operate that heavy equipment you're going to need a keen technical brain, to be good with numbers, comfortable with technical documents, quick to understand complex engineering systems and have the ability rapidly to decode, integrate and evaluate hard data from a wide range of sources in order to make rational and justifiable decisions based on hard criteria.[30]

Commercial pilots are referred to as captains for a reason. They are in command of an airliner, and everyone in it, including the cabin crew and passengers. This requires leadership qualities and good communication skills to clearly convey information to copilots, flight engineers, air traffic controllers, and ground crews. And oftentimes when cabin crews are tired and jet-lagged from long flights, the pilot is required to act as a counselor, according to the pilot who blogs under the name Airline Captain:

> I believe there should be a day on every command course where a therapist is brought in to teach Captains . . . how to deal with people who may be very emotional. We aren't just there to manage the aircraft, we're managing 200 people and 5 crew members. . . . [I have often] felt like I am meant to be playing the role of a therapist. I think tiredness heightens emotions and crew are usually exhausted, so much of it comes out at work.[31]

Workplace Settings

On the job, pilots endure the same tedium at airports as passengers do. They must deal with crowds, overpriced food in airport terminals, and long delays caused by weather and other problems. And just like everyone else, pilots can suffer from jet lag, fatigue caused by flying through different time zones. To avoid exhaustion, the FAA limits pilots to fewer than one hundred hours of flight time per month, although most only fly around seventy-five hours. Preflight and postflight chores on the ground do not count as flight time. Commercial pilots working for major airlines often work twelve-hour days. They might complete several trips, called flight sectors, in a single shift. Rida Rais says, "I have completed 6 flight sectors in a single day, each consisting of a very short 40–50-minute flight. These flights are scheduled respecting the maximum duty period allowed for pilots in a day. Between each sector, there is a 30-minute turnaround time, during which we make necessary preparations for the next flight segment."[32]

Pre- and Post-Takeoff Checklists

"ORD [Chicago O'Hare International Airport] is busy this morning so it takes a few minutes for us to get in line to takeoff. We configure the airplane for takeoff and verify with more checklists to make sure everything is in order, then we get our clearance to take off. We line up on the runway, add full power, and at about 150 mph I pull back on the yoke, and we become airborne. After several heading and altitude changes to get us clear of the congested ORD airspace, and—you guessed it—a few more checklists, we're well on our way."

—Thomas Fogerty, reserve pilot

Thomas Fogerty, "A Day in the Life of a Reserve Pilot," Cleared to Dream, 2025. https://clearedtodream.org.

Most pilots spend a considerable amount of time away from home and family and often complain about lonely nights in hotel rooms.

Employers and Pay

Airline pilots work for well-known passenger airlines like Delta Air Lines, Southwest Airlines, American Airlines, and United Airlines. Cargo pilots work for freight companies that transport mail, packages, and merchandise. Corporate pilots fly business executives or cargo for private companies. Air tour pilots operate sightseeing flights that take tourists through scenic areas. Agricultural pilots spray pesticides and fertilizers on farmland. Medical evacuation pilots usually operate helicopters that transport patients in emergency situations.

The Bureau of Labor Statistics (BLS) says the median annual wage for all pilots and copilots in 2024 was $198,100. However, pilots and copilots who worked for major airlines were earning $226,600, while those who flew for couriers and express delivery services like United Cargo were bringing in $239,200 or more. Those who flew for the federal government were at the lower end of the pay scale, earning $130,980 annually.

What Is the Future Outlook for Commercial Pilots?

The BLS say the job outlook for commercial pilots is expected to grow by 5 percent through 2033. Airports are as busy as ever, and more people are flying every year. While these conditions remain in place, those who can perform the duties of commercial pilots will be in demand.

Find Out More

Aircraft Owners and Pilots Association (AOPA)

www.aopa.org

The AOPA offers general aviation news, and its website provides comprehensive information about flight schools, safety, and careers in aviation. Students can find detailed steps for obtaining a pilot's license, along with information about scholarships, flight training loans, and aircraft purchases.

Air Line Pilots Association, International (ALPA)

https://clearedtodream.org

This association—representing seventy-five thousand pilots—focuses on safety and security issues. The ALPA website provides insight into a commercial pilot's work life and provides "getting started" information for K–8 students and high school students.

International Society of Women Airline Pilots (ISA)

https://isa21.org

The ISA supports women who hold commercial pilot licenses. The ISA website covers the activities of female pilots in books and new stories and provides scholarship opportunities for women seeking a career in aviation.

Source Notes

Introduction: Earning on the Go

1. Caroline Rowe, "How I Make Money from Travel Blogging," *Pack the Suitcases* (blog), June 10, 2024. www.packthesuitcases.com.
2. Shaelyn Combs, "40+ Business Travel Statistics for 2024," Expensify, September 10, 2024. https://use.expensify.com.
3. Alex Walling, "The Realities of Traveling for Work," Medium, October 13, 2018. https://medium.com.
4. Sejal, "The Pros and Cons of Traveling Jobs," *The Lazy Gal* (blog), November 26, 2023. www.thelazygal.com.

English Teacher Abroad

5. Erika Copeland, "How I Made Teaching English a Career," ESL Teacher 365, 2025. www.eslteacher365.com.
6. Copeland, "How I Made Teaching English a Career."
7. Mark Williams, "Visa Nightmares and Work Permit Woes: The Bureaucratic Struggles of TEFL Teachers," Medium, March 31, 2025. https://mark78.medium.com.
8. Emma Lander, "10 Fun ESL Games and Activities for Teaching Kids English Abroad," Go Overseas, May 22, 2024. www.gooverseas.com.
9. William Davies, "The Ultimate Guide to Teaching English in China (2025)," Go Overseas, November 26, 2024. www.gooverseas.com.

Flight Attendant

10. Quoted in Vivian Ho, "How Did Everyone Survive the Delta Plane Crash in Toronto?," *Washington Post*, February 18, 2025. www.washingtonpost.com.
11. Kara Watts, "Pros and Cons of Being a Commercial Flight Attendant," Flight Attendant Life, 2025. https://flightattendantlife.com.

12. Jennifer Pierce, "The Things You Might Not Know About Flight Attendants," Simply Flying, April 8, 2025. https://simpleflying.com.

Construction Manager

13. Quoted in Shae Conner, "Q&A: Professor Explains Why Students Flock to UTSA for Construction Management Careers," UTSA Today, October 16, 2024. www.utsa.edu.
14. Celeste Gable, "A Day in the Life of a Construction Manager: Roles and Responsibilities," Raymond Research Group, October 25, 2023. https://raymondsearchgroup.com.
15. Quoted in Eden, "What Does a Typical Day Look like for a Construction Manager?," CareerVillage.org, March 21, 2023. www.careervillage.org.
16. John Hansen, "How Hard Is the Job of Construction Managers? I'm Getting My Degree and Am Now a Freshman," Quora, 2023. www.quora.com.

Travel Nurse

17. Jennifer Traub, "My Travel Nurse Career: Then and Now," Gypsy Nurse, April 9, 2025. www.thegypsynurse.com.
18. Debra Wood, "A Day in the Life of a Travel Nurse," AMN Healthcare, October 5, 2023. www.amnhealthcare.com.
19. Wood, "A Day in the Life of a Travel Nurse."
20. Scott Barry, "A Typical Day in the Life as a Travel Nurse," Ventura Medstaff, October 2, 2024. https://venturamedstaff.com.
21. Quoted in Wood, "A Day in the Life of a Travel Nurse."
22. Kelita Raphael, "A Day in the Life of a Travel Nurse," ATI, 2025. https://atinursingblog.com.
23. Katie Dixon, "Inside the Travel Nurse Lifestyle," Triage, April 22, 2024. https://triagestaff.com.

Sales Engineer

24. Quoted in Ned Bellavance and Ethan Banks, "Day Two Cloud 186: A Day in the Life of a Sales Engineer with Pete Robertson," Packet Pushers, March 15, 2023. https://packetpushers.net.
25. Andrey Melnichuk, "The Benefits of a Career as a Sales Engineer," Careerist, 2024. www.careerist.com.
26. Yang Lu, "What Will the Regular Daily Job of a Sales Engineer Be Like?," Quora, 2025. www.quora.com.
27. Ramzi Marjaba, "A Week in a Life of a Sales Engineer," We the Sales Engineers, 2025. https://wethesalesengineers.com.
28. Quoted in Bellavance and Banks, "Day Two Cloud 186."

Commercial Pilot

29. Quoted in Karina Moisejenko, "Day in the Life of an Airline Pilot," BAA Training, August 28, 2023. https://baatraining.com.
30. Alec Myers, "Life as an Airline Pilot," Alec Myers Flight Training, July 30, 2023. https://alecmyersflighttraining.com.
31. Airline Captain, "3 Hardest Parts of Being an Airline Captain (That They Don't Teach You on the Command Course!)," *The Diary of an Airline Captain* (blog), July 31, 2024. https://pilotbible.com.
32. Quoted in Moisejenko, "Day in the Life of an Airline Pilot."

Interview with an English Teacher Abroad

Since 2009 Erika Copeland has taught English abroad in six countries, including South Korea, Germany, France, the Republic of Georgia, and Poland. Her current focus is teaching English online through her website English with Reeka. Copeland answered the following questions by email. Her answers have been edited for length and clarity.

Q: Why did you become an English teacher abroad?

A: I wanted to have the experience of living abroad while making money and to develop my résumé for future jobs and career prospects. I graduated from college in December of 2008, and I knew that teaching English abroad was a popular choice for recent college graduates. I applied for a teaching job with the Gyeonggi English Program in South Korea (GEPIK) after a friend recommended it to me. In the fall of 2009, I decided to officially "move abroad." I did not expect this to be my life for the past 17 years, but here we are.

Q: Can you describe a typical workday?

A: For most English teachers abroad, a normal day is like any other day for a teacher working in a school. Some jobs only require you to be at school during your allocated lesson hours, while other schools want you to be at the school all day, whether you are teaching during that time or not. As an online teacher, my typical workday starts in the morning or early afternoon and lasts until 9 pm. I tend to have morning lessons, a big break in the afternoon, and then evening lessons. It has really helped me manage my time better and to be refreshed between lesson block periods.

Q: What do you like most about your job?

A: What I love most about my job is that I am the boss, and I decide which days and times I will work. I can also travel while working if I want to. Of course, meeting different personalities and nationalities from around the world is also the best part. Some of my students have been with me for a long time now, and we have developed very endearing relationships. They trust me as their teacher, and I appreciate that they want to have lessons with me because they have directly chosen me to be their teacher.

Q: What do you like least about your job?

A: As a small online business owner, you cannot rely on a fixed salary. Every month is different. Students come and go, and then you must try to find replacements for them as quickly as possible to make sure you will make a proper income. National holidays, summer holidays, and travel plans by my students can greatly affect my monthly salary. I always know that while living in the Republic of Georgia the months of January and August will be my "poorest" months because these are the two months that people in this country are not available for lessons because they are traveling, enjoying life, and celebrating holidays. I try to always be prepared financially for these two particular months.

Q: What personal qualities do you find most valuable for this type of work?

A: To be an online teacher or a teacher working abroad, you must be someone who can handle being out of your comfort zone for long periods of time. You will always be put into situations that feel absolutely abnormal. Sometimes schools will just throw you into a classroom and expect you to just "teach" without any instruction or materials. It's a bit nutty to be honest. Being able to "think outside the box" or come up with a lesson on a whim will set you apart from the rest, and you may be able to make this job a lasting career, like I have.

Q: What advice do you have for students who might be interested in this career?

A: For anyone interested in this career, my greatest advice is to take TESOL (Teaching English to Speakers of Other Languages) and TEFL (Teaching English as a Foreign Language) courses in real life and not just online. An online TESOL/TEFL course will never prepare you for the classroom environment, which is not for the faint of heart.

Most schools want their native English teachers to have a minimum of a bachelor's degree. If you think you may want to make teaching abroad a real career, then definitely get a teaching degree and apply for international schools around the world. They always have amazing international teacher packages available, including free housing, health insurance, flight tickets, a moving stipend, and competitive salaries. You will feel very differently as a teacher working in an international school than being just a "guest teacher" at let's say a public school or *hagwon* in South Korea (for example). Always do your research thoroughly before signing any contracts and making a big move across the world.

Q: What types of projects are you currently working on?

A: I am continuing to build my social media accounts online, especially on TikTok and Instagram. I am trying a build a following and a community of people who are interested in learning the English language. I also am trying to target other Americans who are looking to start their lives abroad as teachers or in other careers. Building a business from scratch has been very difficult for me at times but definitely worth every minute spent on building something long-lasting. I have manifested a great career through hard work, grit, determination, and nonstop networking.

Other Jobs If You Like Travel

- Adventure tour guide
- Airway transportation systems specialist
- Aviation safety inspector
- Bus driver
- Civil engineer
- Cruise ship director
- Delivery driver
- Diesel service technician
- Event planner
- Excursions manager
- Fashion designer
- Film director
- Fishing and hunting guide
- Flight engineer
- Flight instructor
- Industrial engineering technician
- Journalist
- Machine operator
- Marine biologist
- Mobile equipment service technician
- Musician
- Park ranger
- Photographer
- Public relations manager
- Railroad worker
- Surveyor
- Tour guide
- Tractor trailer driver
- Water transportation worker
- Wildfire fighter

Editor's note: The online *Occupational Outlook Handbook* of the US Department of Labor's Bureau of Labor Statistics is an excellent source of information on jobs in hundreds of career fields, including many of those listed here. The *Occupational Outlook Handbook* may be accessed online at www.bls.gov/ooh.

Index

Picture Credits

Cover: YAKOBCHUK VIACHESLAV/Shutterstock

10: Rido/Shutterstock
19: Hananeko Studio/Shutterstock
26: Ground Picture/Shutterstock
34: Gorodenkoff/Shutterstock
50: Sunshine Seeds/Shutterstock

About the Author

Stuart A. Kallen is the author of more than four hundred nonfiction books for children and young adults. He has written on topics ranging from the theory of relativity to making a career in the music industry. In 2018 Kallen won a Green Earth book award from the Nature Generation environmental organization. He has also written award-winning children's videos and TV scripts. In his spare time, he is a singer, songwriter, and guitarist in San Diego.